COMETS

SETH KINGSTON

PowerKiDS press.
New York

Published in 2021 by The Rosen Publishing Group, Inc.
29 East 21st Street, New York, NY 10010

First Edition

Editor: Elizabeth Krajnik
Book Design: Reann Nye

Photo Credits: Cover solarseven/Shutterstock.com; series art Zoteva/Shutterstock.com; series art Abstractor/Shutterstock.com; p. 5 JOHN THOMAS/SCIENCE PHOTO LIBRARY/Science Photo Library/Getty Images Plus/Getty Images; p. 6 https://en.wikipedia.org/wiki/Edmond_Halley; pp. 7, 9 (top), 14 Vladi333/Shutterstock.com; pp. 9 (bottom), 22 Triff/Shutterstock.com; p. 11 Dneutral Han/Moment/Getty Images; p. 12 Paul Fleet/Shutterstock.com; p. 13 shooarts/Shutterstock.com; p. 15 Time Life Pictures/The LIFE Picture Collection/Getty Images; p. 16 https://commons.wikimedia.org/wiki/File:Bayeux_Tapestry_32-33_comet_Halley_Harold.jpg; p. 17 Getty Images/Hulton Archive/Getty Images; p. 19 (top) Education Images/Universal Images Group/Getty Images; p. 19 (bottom) Manfred_Konrad/iStock/Getty Images Plus/Getty Images; p. 20 VW Pics/Universal Images Group/Getty Images.

Library of Congress Cataloging-in-Publication Data

Names: Kingston, Seth, author.
Title: Comets / Seth Kingston.
Description: New York : PowerKids Press, [2021] | Series: Lighting up the sky | Includes index.
Identifiers: LCCN 2019048273 | ISBN 9781725318373 (paperback) | ISBN 9781725318397 (library binding) | ISBN 9781725318380 | ISBN 9781725318403 (ebook)
Subjects: LCSH: Comets–Juvenile literature.
Classification: LCC QB721.5 .K46 2021 | DDC 523.6–dc23
LC record available at https://lccn.loc.gov/2019048273

Manufactured in the United States of America

Some of the images in this book illustrate individuals who are models. The depictions do not imply actual situations or events.

CPSIA Compliance Information: Batch #CSPK20. For Further Information contact Rosen Publishing, New York, New York at 1-800-237-9932.

CONTENTS

LOOK UP!

If you look up at any given moment, you might see an impressively bright comet. Comets—like planets, moons, and stars—are **celestial** bodies. They travel millions of miles before we see them as they move closer to the sun or to Earth. And some put on quite a light show!

Even though many comets come unannounced and aren't visible to the naked eye, some comets have a lasting effect. In March 1996, Comet Hyakutake came within 9.3 million miles (15 million km) of Earth. It reached a **magnitude** of zero, which means it was very bright, and its tail reached across more than half the sky.

Comet Hyakutake made its closest approach to Earth on March 26, 1996. At that time, it was the brightest comet people had seen in 20 years.

UNDERSTANDING COMETS

For thousands of years, scientists didn't understand comets. Aristotle, a **philosopher** who lived in ancient Greece, thought comets were objects Earth breathed out that caught fire in the **atmosphere**. This idea stuck until 1577, when Danish **astronomer** Tycho Brahe found that comets weren't in Earth's atmosphere but were even farther away from Earth than the moon.

SEEING THE LIGHT

English astronomer Edmond Halley used Newton's math and found that three comets had similar orbits and that they might actually be one comet that returns every 76 years. We know it today as Halley's Comet.

In the late 1600s, scientist Isaac Newton used math to show that comets orbit in space. Some comets have parabolic, or half-oval-shaped, orbits and won't ever be seen again. Others have elliptical, or oval-shaped, orbits and will be seen again.

PARTS OF A COMET

A comet is made up of a nucleus, coma, and one or more tails. The nucleus is made of ice, dust, and other matter. It's the heart of the comet and the only part that always exists.

The coma is a cloud of dust, gas, and water vapor around the nucleus. The tail is the part that extends from the comet and points away from the sun. A comet's coma and tail form as the comet nears the sun. A comet may have a dust tail or a gas **ion** tail. Each type of tail forms under different conditions and is a different color.

SEEING THE LIGHT

"Comet" comes from the Greek word *kometes*, which means a head with long hair.

If a comet's tail appears blue, that means it's a gas ion tail. Dust tails are usually yellow. Comets can have both types of tails or even no tail at all.

WHAT'S THE DIFFERENCE?

Asteroids, meteoroids, and comets are similar to each other, but have key differences. Asteroids are big pieces of rock orbiting in the solar system. They come from the asteroid belt between Mars and Jupiter. Asteroids don't usually have tails, even when they're close to the sun. Asteroids have shorter, more circular orbits and often group together in belts.

Meteoroids are smaller than asteroids and are made of rock and other matter from space. When a meteoroid enters Earth's atmosphere, it's called a meteor. A meteorite is a meteor that doesn't burn up in the atmosphere and hits Earth. Meteors are also called shooting stars.

Many people enjoy gathering to watch meteor showers. To watch a meteor shower, you should go somewhere very dark and lie on the ground facing toward the sky. Meteor showers are best seen later in the evening.

WHERE DO COMETS COME FROM?

Comets come from two places in our solar system. Scientists think short-period comets, those that orbit the sun in less than 200 years, come from the Kuiper belt. The Kuiper belt is shaped like a disc and contains hundreds of millions of comets and other small celestial bodies. It exists beyond Neptune's orbit, about 30 to 100 astronomical units (AU) away from the sun. One AU equals 93 million miles (150 million km).

SEEING THE LIGHT

Scientists think the Kuiper belt is made up of objects left over from the formation of the solar system about 4.6 billion years ago.

Comets that come from the Kuiper belt orbit in the same direction as the planets in our solar system. The Oort cloud is where new comets come from.

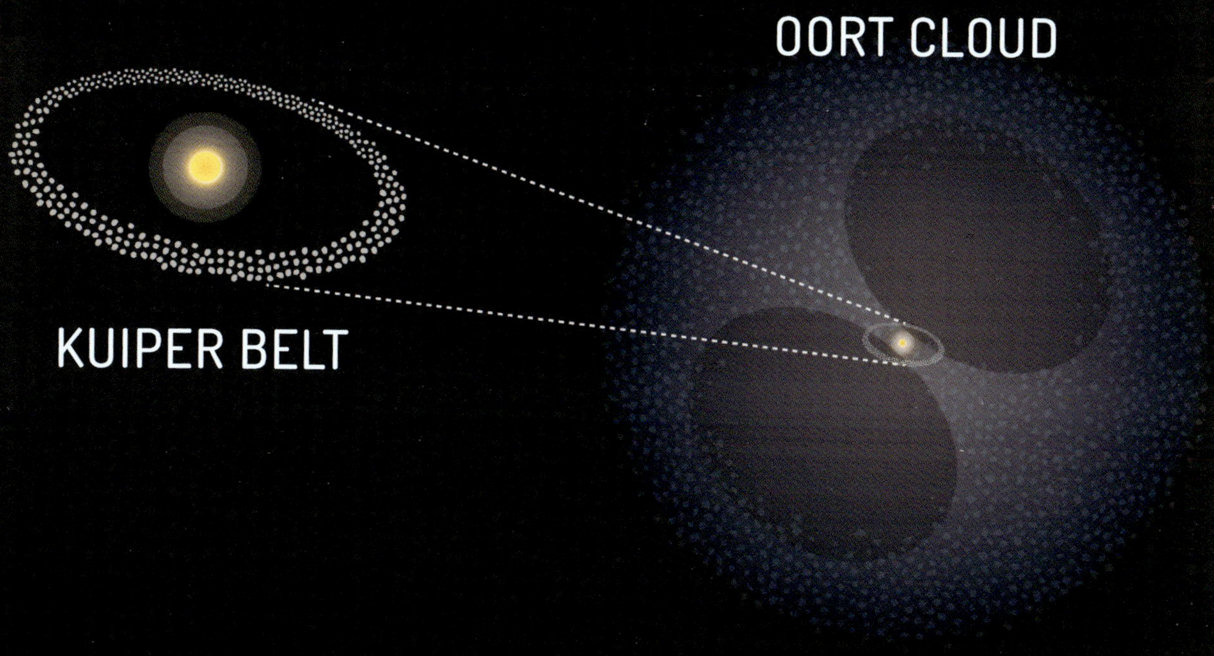

OORT CLOUD

KUIPER BELT

Long-period comets, those that orbit the sun in more than 200 years, come from the Oort cloud, which surrounds the solar system like a giant bubble. The Oort cloud is the most distant part of our solar system. Astronomers believe it's 50,000 AU from the sun.

HOW DO COMETS MOVE?

Comets got started on their orbit with energy from the creation of the solar system. Comets, like planets, move by being pulled by the gravity of planets and even stars. However, comets' orbits aren't as fixed as planets' orbits. The heat and light from the sun power the comet.

SEEING THE LIGHT

Comets' tails are always pointed away from the sun. The tails become longer the closer the comet gets to the sun.

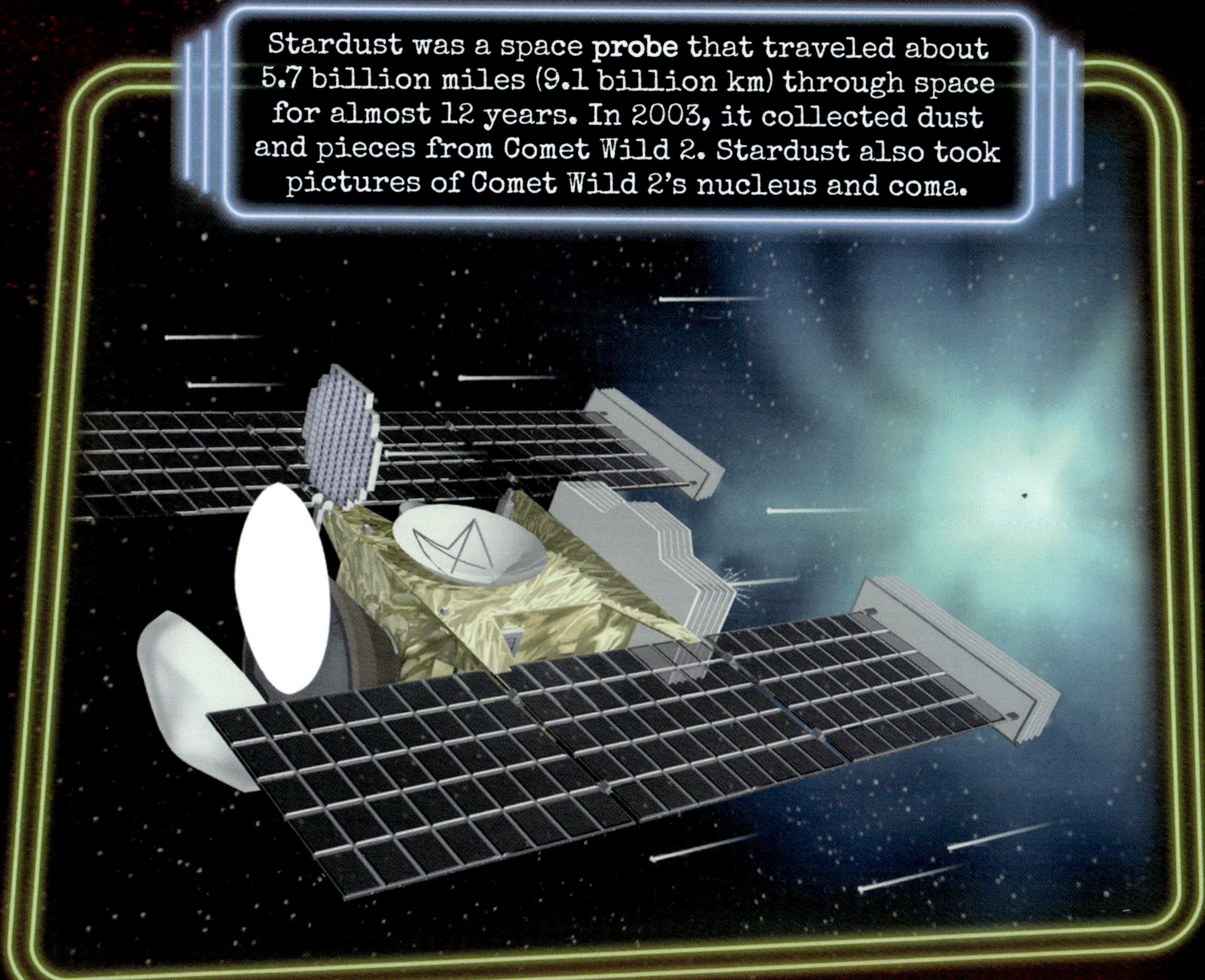

Stardust was a space **probe** that traveled about 5.7 billion miles (9.1 billion km) through space for almost 12 years. In 2003, it collected dust and pieces from Comet Wild 2. Stardust also took pictures of Comet Wild 2's nucleus and coma.

When a comet is closest to the sun, the comet is at perihelion—*peri* meaning "near" and *helios* meaning "sun." When a comet is at the point of its orbit farthest from the sun, the comet is at aphelion—*apo* meaning "away from" and *helios* meaning "sun."

1P/HALLEY

1P/Halley, or Halley's Comet, is one of the most well-known comets. In 1705, Edmond Halley had been studying the orbits of 24 comets. He saw that comets in 1531, 1607, and 1682 had similar orbits. He thought these three comets were actually one comet that passed Earth every 76 years. Halley died in 1742, and the comet returned during 1758 and 1759. The comet was named after him.

SEEING THE LIGHT

The Bayeux **Tapestry**, which shows the Norman **Conquest** of England in 1066, has a scene that shows people looking up at a bright star. Halley's Comet appeared six months before the Norman Conquest.

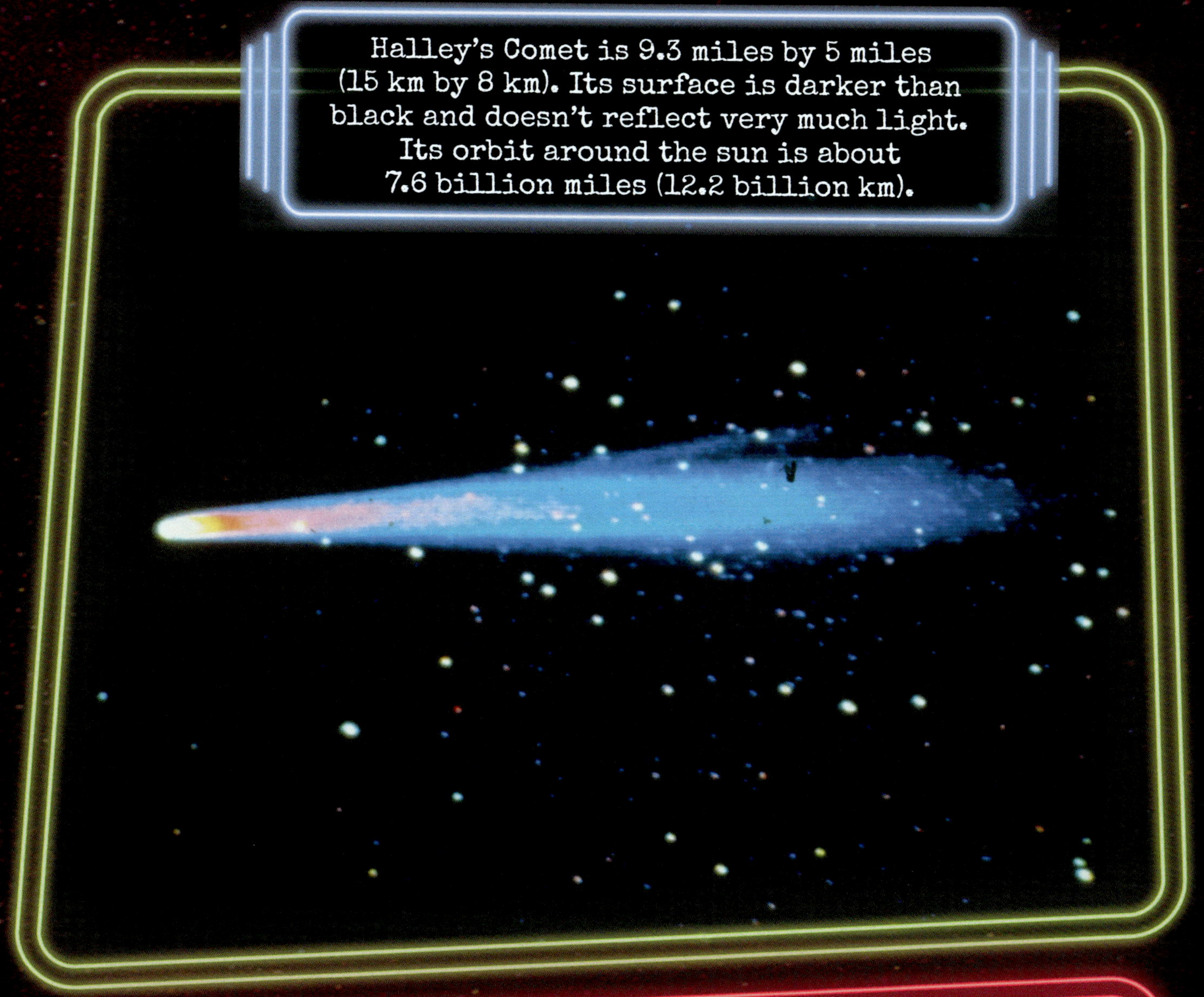

Halley's Comet is 9.3 miles by 5 miles (15 km by 8 km). Its surface is darker than black and doesn't reflect very much light. Its orbit around the sun is about 7.6 billion miles (12.2 billion km).

During Halley's Comet's most recent appearance during 1986 and 1987, scientists photographed the comet for the first time. Seeing Halley's Comet close up helped scientists understand what the comet is made of.

COMETS HALE-BOPP AND HOLMES

On July 23, 1995, Alan Hale and Thomas Bopp discovered a very bright comet outside of Jupiter's orbit. This comet was brighter than Halley's Comet would be at the same distance. It was the brightest comet since Comet West in 1976. Comet Hale-Bopp is about 37 miles (60 km) across.

Sometimes, comets' nuclei display outbursts, or sudden releases of dust and gas. Although Comet Holmes was discovered in 1892, it was hard to see without a telescope. In October 2007, Comet Holmes's nucleus had an outburst, making it 1 million times brighter. Scientists are unsure why Comet Holmes had such a bright outburst.

SEEING THE LIGHT

Comet Holmes's coma grew to be more than 869,900 miles (1.4 million km) across. This is bigger than the sun! Before its outburst, Comet Holmes's coma was just 2.1 miles (3.4 km) across.

Some scientists think that Comet Holmes may have released 100 million tons (90.7 million t) of dust into space.

NAMING COMETS

Until 1994, comets were designated by the year they were discovered and a lowercase letter that stood for the order in the year they were discovered. For example, the first comet discovered in 1994 would be 1994a.

Comet C/2013 R1 (Lovejoy) is the designation for the comet that has a nonperiodic orbit, was discovered in 2013, was the first comet discovered in the first half of September, and was discovered by Terry Lovejoy.

P/	COMET ON A KNOWN, PERIODIC ORBIT
C/	COMET ON A NONPERIODIC ORBIT
X/	COMET WITH AN UNKNOWN ORBIT
D/	PERIODIC COMET THAT HAS DISAPPEARED, BROKEN UP, OR BEEN LOST

Today, a comet has a letter that tells what type of comet it is, followed by the year of its discovery, a letter that tells in which half of the month it was discovered, and a number that tells what number of comets have been discovered in that half month. Sometimes the name of the person, people, or observatory that discovered the comet will follow.

TELESCOPE TIME!

Astronomers usually let people know when a comet is coming. Other astronomers have a list of expected comets, which month of which year they will be visible, which part of the day they'll be visible during, how bright they are, and other useful facts.

If you're interested in seeing comets, you may want to buy a telescope. Most comets aren't visible to the naked eye. Your best chance of seeing a scheduled comet is if you go to a place without streetlights or other lights, such as a state or national park or a dark sky reserve. You could even discover a comet and have it named after you!

GLOSSARY

astronomer: A person who studies astronomy, which is the scientific study of stars, planets, and other objects in outer space.

atmosphere: The whole mass of air that surrounds Earth and other celestial bodies.

celestial: Of or relating to the sky.

conquest: The act or process of taking control of a country, city, or group of people by force.

ion: A small bit of matter that has an electrical charge.

magnitude: A number that shows the brightness of a star.

philosopher: A person who tries to discover and to understand the basic nature of knowledge.

probe: An unmanned craft used to send facts about a body in space.

tapestry: A heavy cloth that has pictures woven into it and is often used as a wall hanging.

Due to the changing nature of Internet links, PowerKids Press has developed an online list of websites related to the subject of this book. This site is updated regularly. Please use this link to access the list:
www.powerkidslinks.com/lus/comets